Dmitry Khrustalev
Marat Ibrayev
Olga Tyagunova

Production of phenol-formaldehyde foam under MVA conditions

Dmitry Khrustalev
Marat Ibrayev
Olga Tyagunova

Production of phenol-formaldehyde foam under MVA conditions

Step-by-step instructions for making phenol-formaldehyde foam under microwave activation conditions

ScienciaScripts

Imprint
Any brand names and product names mentioned in this book are subject to trademark, brand or patent protection and are trademarks or registered trademarks of their respective holders. The use of brand names, product names, common names, trade names, product descriptions etc. even without a particular marking in this work is in no way to be construed to mean that such names may be regarded as unrestricted in respect of trademark and brand protection legislation and could thus be used by anyone.

Cover image: www.ingimage.com

This book is a translation from the original published under ISBN 978-3-330-31904-2.

Publisher:
Sciencia Scripts
is a trademark of
Dodo Books Indian Ocean Ltd. and OmniScriptum S.R.L publishing group

120 High Road, East Finchley, London, N2 9ED, United Kingdom
Str. Armeneasca 28/1, office 1, Chisinau MD-2012, Republic of Moldova, Europe
Managing Directors: Ieva Konstantinova, Victoria Ursu
info@omniscriptum.com

Printed at: see last page
ISBN: 978-620-7-43715-3

TABLE OF CONTENTS

INTRODUCTION

On May 30, 2013 the President of the Republic of Kazakhstan signed a concept on the transition of the Republic of Kazakhstan to a "green economy". One of the most important tasks of this program is the transition to resource-saving and energy-efficient technologies, which is fully consistent with global trends in the development of modern science and society as a whole. The introduction of "green technologies" will allow the Republic of Kazakhstan to enter the 30 most economically developed countries of the world by 2050. Introduction of conceptually new, "green" chemical production is the most important component of this program [1].

The research described in this book is devoted to a new method for the production of phenol-formaldehyde foams. These materials have been known since the middle of the 20th century. Their undoubted advantages include mechanical strength and chemical resistance. The temperature range of operation of these polymers is extremely wide and ranges from -200° C to $+200^\circ$ C. Phenol-formaldehyde foams, due to their unique properties could find the widest application in various fields of technology. Their application for thermal insulation of various construction objects, in particular for roofs, walls, pipes is very promising due to low thermal conductivity combined with excellent fire resistance, which are inherent in these materials. However, the existing methods (mechanical foaming of aqueous dispersions; foaming of partially cured compositions) require the use of appropriate hardeners, solvents, catalysts, gas formers, which is a significant disadvantage of all classical methods. In addition, classical methods require large amounts of water, electricity and a number of auxiliary substances, which are not included in the final product and pollute the environment. It is for these reasons that phenol-formaldehyde foams have received limited application only in high-budget projects, for example, in the construction of nuclear power plants.

The peculiar nature of microwave irradiation made it possible to bypass the above problems and develop a highly efficient, environmentally friendly, economically cost-effective method for the synthesis of phenol-formaldehyde foam. Using methods of mathematical planning of the experiment it was found that the desired product under conditions of microwave activation can be successfully produced within 5 minutes, while in traditional conditions this process takes at least 20 hours.

This book will be useful for students, master's students, doctoral students, researchers and anyone interested in the possibilities of organic synthesis under microwave activation conditions.

Authors

The research was carried out under the grant "Development of innovative technologies of production of industrially demanded composite materials in conditions of microwave activation" (2015) of the Ministry of Education and Science of the Republic of Kazakhstan Scientific supervisor - Dr. Khrustalev D.P. Khrustalev.

Project team

Dmitry Khrustalev currently works as an associate professor at the Department of Pharmaceutical Disciplines and Chemistry of Karaganda State Medical University. In 2010 he defended his doctoral dissertation on the topic: "Synthesis and modification of industrially demanded nitrogen-containing heterocyclic compounds under microwave irradiation". In 2010 he was awarded the Prize of the Fund of the First President of the Republic of Kazakhstan, in 2011 the Prize "Patriot of the Year" for research in the field of microwave chemistry. Author of more than 150 scientific papers.

Marat Ibrayev - Doctor of Chemical Sciences, Professor of the Department of "Industrial Ecology and Chemistry" at Karaganda State Technical University. For more than ten years he has been implementing projects on fine organic synthesis of biologically active substances, modification of natural compounds, complex processing of coal and technogenic wastes. Author of more than 150 scientific papers.

Vladimir Fedorchenko - PhD in Technical Sciences, head and executor of a number of scientific projects on complex processing of industrial waste and synthesis of composite materials. Author of more than 100 scientific papers.

Olga Tyagunova - Bachelor of Chemistry and Technology, performer of a number of scientific projects on complex processing of industrial waste and synthesis of composite materials by methods of microwave chemistry. Co-author of a number of scientific articles. Executor of 5 projects of the Ministry of Education and Science of the Republic of Kazakhstan.

Anastasia Khrustaleva Bachelor of Chemistry and Technology, performer of a number of scientific projects on organic synthesis under microwave activation conditions. Winner of the contest of productive innovations of the National Agency for Technological Development. Co-author of a number of scientific articles.

Sanakulova Saniya Master of Chemistry and Technology, performer of a number of scientific projects on organic synthesis under conditions of microwave activation and complex processing of industrial waste. Co-author of a number of scientific articles.

1 PECULIARITIES OF MICROWAVE SYNTHESIS. PRODUCTION OF PHENOL-FORMALDEHYDE FOAMS UNDER CONVECTION HEATING CONDITIONS

1.1 Features and differences between microwave-activated heating and convection heating.

Only in 1986, three groups of researchers led by Robert Gedye (Laurentian University in Ontario, Canada), George Majetich (University of Georgia, USA) and Raymond Giguerce (University of Mercer, USA) attempted to use microwave radiation in chemical synthesis. They discovered that many chemical reactions proceeded thousands of times faster under microwave irradiation compared to conventional thermal heating. After these groundbreaking studies, the interest of the scientific community in studying the effects of microwave heating on organic reactions increased dramatically. These studies have found that organic reactions under microwave heating conditions have significant advantages over conventional activation methods. To date, microwave chemistry has actually developed into an independent discipline [2].

Microwave radiation is electromagnetic radiation with frequencies in the range from 0.3 to 300 GHz, corresponding to wavelengths from 1 cm to 1 m. The microwave region of the electromagnetic spectrum lies between the infrared and radio frequency regions of the electromagnetic spectrum. The microwave radiation range between 1 cm and 25 cm is used extensively in technology for radar and telecommunication devices.

The terms "microwaves", "microwave radiation", "microwave irradiation" are borrowed from foreign literature and are used along with the more established in the CIS countries term "ultra-high frequency radiation" or "UHF", which defines the same frequency range. The terms "Microwave Synthesis", "Microwave Chemistry" are widely used in foreign literature to denote organic or inorganic reactions, processes carried out under microwave irradiation.

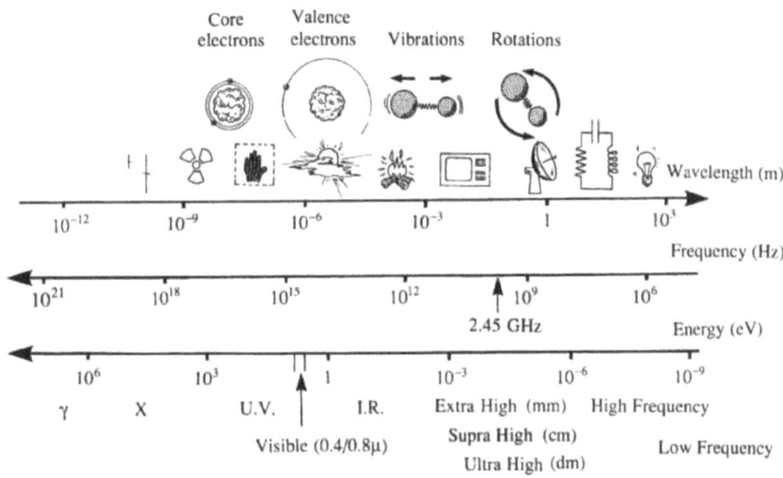

Figure 1.1 - Wavelength to energy ratio in the radio frequency wavelength range

In order to avoid interference with communication devices, international agreements have been developed, according to which several frequencies have been defined for industrial, scientific and medical purposes (Industrial, Scientific and Medical frequencies). For the operation of household microwave ovens, as well as for industrial microwave reactors, the frequency 2.45 GHz has been defined. In addition to the 2.45 GHz frequency, four other radio frequencies in the microwave spectrum band (433 MHz, 915 MHz, 5800 MHz, 24125 MHz) have been allocated for the operation of microwave heating devices.

Traditionally, organic reactions are carried out by convection heating (e.g. oil, water bath or Wood's alloy), where heat is transferred from the heat source to the walls of the reaction vessel and from them to the reaction mass. This is a slow and inefficient way of energy transfer into the system, because it depends on the thermal conductivity and thermal resistance of the material from which the reaction vessel is made, since the temperature of the reaction vessel during heating is much higher than the temperature of the reaction mass. In conditions of convection heating energy is transferred from the walls of the vessel to the reaction mass and further spread by diffusion and heat transfer, which determines the presence of a small content of molecules in the reaction mixture, the energies of which are sufficient for the chemical reactions. The depth of chemical reactions is influenced by viscosity, heat capacity of both solvent and reactants. Widely observed in organic chemistry, the processes of osmolization of the reaction mixture are caused by local overheating of the reaction mass.

The method of conducting organic reactions under MB-irradiation conditions eliminates the use of solvents, provides higher yields, reduces reaction time, increases reaction rate to a large extent, improves selectivity and simplifies the processing of reaction mixtures.

The advantages of reactions under microwave irradiation include the following: uniformity of heating of the reaction mass over the entire volume; inertia-free heating, high dynamics of temperature control; absence of coolant; possibility of instantaneous termination of energy supply in case of necessity, etc.

As can be seen from the figure, heating under MV irradiation conditions, starting from the inside, results in uniform "volumetric heating" of the entire reaction mass.

As can be seen from Figure 1.2, heating under MB irradiation conditions, starting from the inside, results in uniform "volumetric heating" of the entire reaction mass. In the case of heating a test tube on an oil bath, the walls of the vessel are heated first, then the heat is transferred to the reaction mass [3].

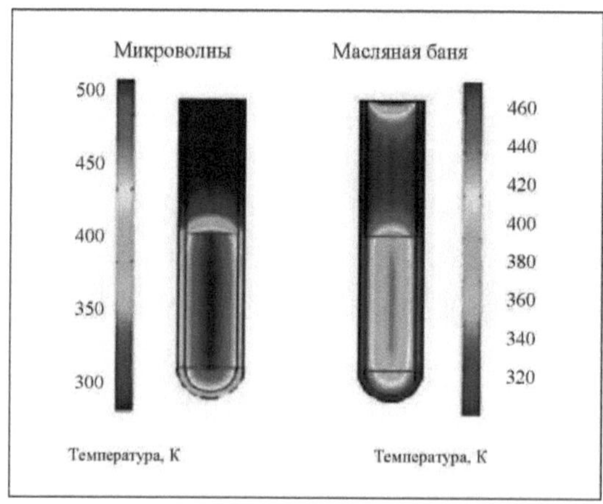

Figure 1.2 - Temperature profile of tube heating process in microwave reactor (left) and on oil bath (right)

Figure 1.3 shows the difference in the heating rate of olive oil in a multimode microwave reactor and an open oil bath [3].

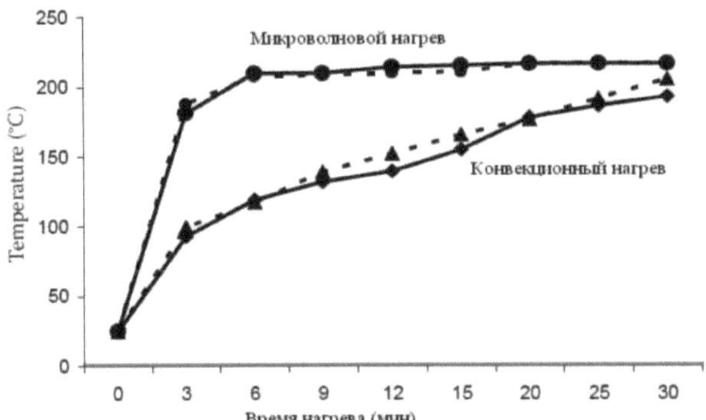

Figure 1.3 - Temperature profile during heating in a closed monomode microwave reactor and in an open oil bath

As follows from the graph in Figure 1.3, dielectric heating under the action of microwave radiation is much more efficient. Thus, under MHE conditions, olive oil is heated to $200°$ C within 5 minutes. Whereas in convection heating conditions, after 5 minutes the temperature reaches only $120°$ C. After 30 minutes, the heating temperature of olive oil under convection heating conditions reaches only $165-170°$ C. Both experiments were performed in the same reaction vessel equipped with an internal temperature sensor. This example clearly shows the advantages of microwave heating.

Thus, the advantages of synthesis under MBI conditions are:

- higher reaction temperatures can be achieved by rapid microwave radiation in a closed autoclave vessel;

- in many cases, the reaction time was significantly reduced, higher yields and purity of reaction products were recorded, thereby increasing the optimization of the process;

- solvents with lower boiling points can be heated to temperatures much higher than their boiling point under pressure (in closed vessels) under MB radiation;

- microwave heating allows the reaction mixture to be heated from the inside, resulting in faster heating;

- there is a so-called microwave radiation effect that cannot be reproduced by conventional heating, e.g., selective synthesis using microwave-absorbing catalysts;

- accessible control of system temperature and pressure;

- microwave heating provides more efficient energy than a classic oil bath heated by direct molecular heating;

- method can be easily adapted to automated serial or parallel synthesis.

The electric component of the electromagnetic field is the cause of heating by three known mechanisms: ionic conduction, dipole polarization, and inter-surface interaction. The factor determining the possibility of conversion of microwave radiation into thermal energy is the presence of a dipole moment in the irradiated molecule. In this case, the polar molecule or ion begins to make oscillatory movements with the frequency of the applied electromagnetic field.

The fundamental mechanism of microwave heating involves oscillations of a polar molecule or ion occurring in an alternating electromagnetic field. The frequency of these oscillations depends on the frequency of the electromagnetic field and can reach 1 billion times per second. Thus, electromagnetic oscillations lead to mechanical friction between particles, which in turn leads to the generation of heat energy.

Different materials interact with microwave radiation in different ways. Some, such as sulfur, is a conductor of microwave radiation, copper reflects microwave radiation, and water absorbs microwave radiation, transforming electromagnetic radiation into heat energy [4].

In the first approximation, only materials absorbing microwave radiation can enter into chemical reactions in the microwave field. In works [5-7] the basic principles of interaction of the MWI with matter are stated. Among the physical phenomena occurring in the interaction of MWI with matter, two effects mainly lead to heat generation: orientational polarization of dipoles and ionic conduction. Some authors add one more: inter-surface mechanism [8].

The dipolar (orientation-polarization) mechanism (Dipolar Polarization) consists in the fact that under the action of the electric component of an alternating electromagnetic field, molecules of polar substances are oriented so that the vectors of their dipole moments are antiparallel to the field lines. In the gas phase, the orientation of molecules is easy, but in the liquid phase, due to the high density of the medium, the orientation is complicated by the processes of intermolecular interaction (friction), accompanied by the absorption of microwave radiation and depends on the frequency of the applied field and the viscosity of the solution. At low field frequency, reorientation of the molecule occurs in the phase of field oscillations, and the liquid is heated insignificantly. At high field frequency, the molecules do not have time to reorient, they do not move, and the radiation energy is not converted into heat. In the microwave range (e.g., at 2.45 GHz), the dipole molecules react to the applied field and begin to rotate, but due to the phase mismatch between the field oscillation (1×10^9) and the dipole rotation (1×10^5), the microwave radiation energy (MRI) is converted into the kinetic energy of the molecules and the solution is heated.

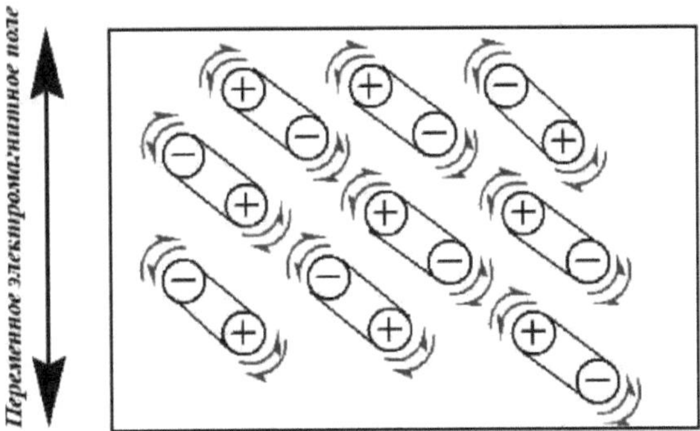

Figure 1.4 - Orientation of dipoles in space

Ion conduction (Conduction Mechanism) [9] is generally similar to the dipole mechanism with the difference that the oscillations are performed by ions.

Interfacial Polarization [10]. This mechanism is a combination of the two previous ones. For such a system, two components are important: a component that transmits microwave radiation and a component that reflects microwave radiation. For example, sulfur is a conductor of microwave radiation and a metal, such as iron, reflects most of the microwave radiation. However, a mixture of these two components absorbs microwave energy well, leading to the desired reaction product. It is important that the substances are used in a finely dispersed form. This is due to the fact that metal powder, unlike sheet metal, absorbs microwave radiation well and the metal particles, when heated, transfer thermal energy into the reaction mixture, which leads to the chemical reaction.

By now it has been established that multiple acceleration of chemical reactions (in hundreds, thousands and in some cases tens of thousands times) occurs due to the influence of one or all three effects: thermal effect (TE), specific thermal microwave (STE) effect, specific non-thermal microwave effect (SNTE). These effects are complex, unusually interesting and voluminous in presentation, the influence of these effects, as well as the dependence of the reaction direction on the type of activation can be found in detail in the book [11].

1.2 . Description of a standard method for the synthesis of phenol-formaldehyde foam plastic

The use of phenol-formaldehyde foams for thermal insulation of various construction objects, in particular for roofs, walls, pipes is very promising due to low thermal conductivity combined with excellent fire resistance, which are inherent in these materials. However, mass production of these materials is constrained by high production costs. This is primarily due to the peculiarities of production, requiring complex hardware design in combination with energy-consuming and environmentally unfriendly technology, which does not allow phenol-formaldehyde foams to compete with cheap and available polystyrene foam. It is for these reasons that phenol-formaldehyde foams have received limited use only in high-budget projects, such as the construction of nuclear power plants [12].

Phenol formaldehyde porous polymers (phenol formaldehyde foams) are derived from phenol formaldehyde resins. By now, phenolic foams are known to have the best fire safety rating of any known foam type used in insulation. Phenolic foams do not burn even when in contact with a blowtorch flame, can withstand heat up to 20-250° C and emit minimal amounts of gases when in contact with a flame.

Despite these technical advantages and generally favorable economic conditions, phenol-formaldehyde foams have not yet penetrated the wide market of thermal insulation. And, due to the high cost of their production, they are used in the manufacture of expensive products: space rockets, airplanes, thermal insulators of nuclear power plants. And their replacement of flammable polystyrene foam will be possible only after reduction of production costs.

Phenolic foams are a capricious material. One of the main reasons why phenolic foams have not gained market acceptance has been the unsatisfactory initial thermal conductivity and the undesirable increase in thermal conductivity over time, which is directly related to production. In addition, performance such as compressive strength of known phenolic foams is not as high as desired for normal use. Existing phenolic foams are known to be brittle and fragile. But their use in composite materials allows to level out the described disadvantages.

The general composition and method of producing phenol-formaldehyde foam is well known. Typically, phenol formaldehyde foams are obtained by mixing resol (previously synthesized from phenol and formaldehyde), a foaming agent, a surfactant, optional additives and acid are added depending on the need to impart certain properties. The curing catalyst is added in amounts sufficient to initiate the curing reaction, which is highly exothermic. The release of heat from the curing reaction causes the blowing agent to evaporate and form the necessary pores.

The foaming process is usually carried out in a closed mold.

A general method for the continuous production of phenolic foams comprises the following: a foaming composition for producing a phenol-formaldehyde polymer according to the resin type, prepared for continuous feeding in a suitable mixing device phenol, formaldehyde, blowing agent, surfactant, optional additives and an acid curing catalyst. The ratio of these components varies depending on the density, thickness and other requirements of the final product.

Phenolic polystyrene is produced using a rather complex technique, strictly following the regulations for obtaining the material. For this purpose, the device applies the listed ingredients uniformly to a moving substrate, which usually has a protective coating, such as cardboard, to which the phenol-formaldehyde foam adheres. The foam-coated composition is then transferred to a press apparatus where the curing process takes place with the release of heat, which causes foaming. The foaming agent evaporates and the resulting phenol polymer mass hardens.

As mentioned earlier, one of the major disadvantages of phenolic foams is poor initial thermal conductivity. One of the main reasons is believed to be conduction due to cell wall rupture during foaming and curing. This rupture leads to immediate loss of foaming agent, resulting in poor initial thermal conductivity. Torn cell walls also promote the passage of water into the foam, which also leads to further increases in thermal conductivity. Broken cell walls are believed to have a poor effect on the mechanical compressive strength and other properties of phenolic foams.

The other main reason for the initial poor thermal conductivity in phenolic foams is the loss of blowing agent before the cell walls have had time to sufficiently form.

Another significant disadvantage leading to an increase in thermal conductivity over time. It is believed that the main reasons leading to an increase in thermal conductivity over time are: firstly, the presence of small perforations or pores in the cell walls, including spacers within the polymer from which the cell walls are formed, are connected to each other. These small perforations allow the blowing agent to diffuse over time out of the polymer and, be replaced by air. This slow convection of blowing agent to air results in increased thermal conductivity and loss of thermal insulation. Small perforations also allow phenolic foams to absorb water, thus further increasing thermal conductivity. The perforations are believed to be caused by water that is present in some parts of the foamed phenol-formaldehyde compositions, in particular the catalyst dissolved in the water. One way to avoid the use of water is to use anhydrous organic acids, such as arylsulfonic acids.

Another major cause of loss of thermal conductivity over time is cell wall cracking. Often porous phenol formaldehyde resins have very thin cell walls. When phenolic foam that has thin walls is exposed to high temperatures, the cell walls dry out and crack. In addition, the insulation is quite often subjected to heating and cooling cycles with corresponding expansion and contraction of the thin cell wall, resulting in cracking. The cracking of the thin wall allows the blowing agent that fills the cells to evaporate from the phenolic foams over time, resulting in an increase in thermal conductivity and thus with the loss of thermal insulation.

There are several ways to synthesize phenolic foams in such a way as to keep the walls intact and retain the thermal insulation properties of the polymer. However, we will not review them. As will be shown later, organic substances including chlorine, fluorine-containing substances with boiling points ranging from 5 to 55° C have been used as blowing agents. These blowing agents are classified as undesirable substances (from a Green Chemistry perspective) and their use as a component of thermal insulation is initially a very bad idea. Despite the fact that the thermal conductivity of porous phenol-formaldehyde resin is more than twice as high when the bubbles are filled with air instead of fluorinated hydrocarbons, it is absolutely unacceptable to use the latter from the point of view of modern "green technology". Sooner or later these materials will degrade and all these toxic components will get into the environment.

Other attempts to improve the thermal conductivity of phenolic foams were based on the development of specially modified phenol-formaldehyde resins, the use of surfactants, and various additives to the foamed compositions. None of these methods was commercially successful [13-16].

Phenol-formaldehyde foams are not produced directly from phenol and formaldehyde. A pre-synthesized resin with a molecular weight of 350 to 1500 is introduced into the reaction. The degree of polycondensation is preferred on the basis of the necessary

characteristics of the final product, but usually the molecular preferred weight of resol is 950-1500 a.u.m. Different phenol to formaldehyde ratios are found in the literature to produce phenol formaldehyde foams with good performance properties. Typically, the molar ratio of formaldehyde to phenol ranges from 1.7:1 to 2.3:1. A ratio of 2:1 is considered to be most preferred.

Improved characteristics of phenol and formaldehyde based aqueous resol are obtained by reacting phenol and formaldehyde in desired molar ratios in the presence of a basic catalyst until, as a result, the desired resol has the desired molecular weight and dispersion characteristics. For example, phenol, formaldehyde and catalyst may be loaded into the reactor in desired ratios and the reaction will be carried out to the desired molecular weight of the resol. Alternatively, one or two of the ingredients may be loaded into the reactor and the remaining ingredients added to the reaction mixture over time. Any variations are possible at the manufacturer's discretion, the method of producing the phenolic resin is not critical provided, most importantly, that the phenol and formaldehyde condensation product has the desired molecular weight and dispersibility.

As mentioned previously, the resol should have a molar ratio of formaldehyde to phenol of from about 1.7: 1 to about 2.3: 1. If the ratio is higher than 2.3:1, the resulting phenolic foam may have a residual free formaldehyde content. In addition, if the molar ratio of formaldehyde: phenol exceeds 2.3:1, a resol that has too high a viscosity is formed. Phenol formaldehyde foams prepared from resols having a molar ratio greater than 2.3:1 tend to be too loose and have low compressive strength. If the molar ratio is less than 1.7:1, resols with too low a viscosity are formed, resulting in resins with too thin cell walls. Phenolic resolves having molar ratios less than 1.7:1 are too exothermic to react, making it difficult to trap the foaming agent and keep the cell walls from rupturing. Phenol-formaldehyde foams made from these resols give strong shrinkage.

The phenolic resol should have an average molecular weight, from about 950 to about 1500. If the weight average molecular weight is less than about 800, the resin resins are too reactive and not viscous enough. Phenolic resols with an average molecular weight of less than about 800 react too exothermically, causing cell wall failure and excessive evaporation of the blowing agent. As a consequence, many of the cell walls will be destroyed before cells containing blowing agent are formed.

In addition, phenolic resins of molar weights less than 800, are unable to form strong, thick cell walls. The phenolic resin slumps at the beginning of cure, forming cell walls that are thin. Thin cell walls are easily ruptured by the foaming agent and tend to crack upon drying and during use.

Resols having molecular weights above 1500 tend to have a very viscous consistency and are rather difficult to work with. Nevertheless, they can be used in the preparation of acceptable phenol-formaldehyde foams. Lower molecular weight resins can be diluted with them.

Phenolic resols typically contain free formaldehyde up to 710% and free phenol about 5-7% by weight of the resol. Too much free formaldehyde gives the polymer an unpleasant odor. In addition, free formaldehyde and phenol affect the reactivity and viscosity of the resol and the foaming composition.

Phenolic resins suitable for making foams have viscosities in the range of 6000 cP to 10000 cP at 16% water and 25°C. Viscosity is not a critical factor provided the molar ratios, molecular weights and dispersities are as set forth above.

In obtaining resol, up to 10% of phenol can be replaced by other hydroxybenzoles: cresols, xylenols, ethylphenols, resorcinols, etc.

Similarly, up to 10% of the formaldehyde may be replaced with other suitable aldehydes: glyoxal, acetaldehyde, chloral, furfural and benzaldehyde.

The phenol reagent is added to the reactor, typically as an aqueous solution. The phenol concentration can range from about 50 weight percent to about 95 weight percent. Solutions containing less than about 50 weight percent can be used, but the resulting reaction mixture is obtained very dilute and therefore increases the time to obtain a resol with the desired molecular weight. Alternatively, pure phenol can be used. However, no advantage is obtained using pure phenol over aqueous solutions of phenolic concentrates having a phenolic content of more than about 85 weight percent.

Formaldehyde reagent is added to the condensation reactions as an ingredient at a concentration of from about 30 to about 97 weight percent. Solutions containing less than about 30 weight percent formaldehyde can be used, but, similarly, the time to synthesize resols of the desired molecular weight increases dramatically. Often, paraformaldehyde can be successfully utilized as a source of formaldehyde.

Condensation of phenol and formaldehyde, catalyzed by bases. Typically, alkali and alkaline earth metal based catalysts, carbonates, bicarbonates or oxides are used. However, other basic compounds may be used. Examples of suitable catalysts are lithium hydroxide, sodium hydroxide, potassium hydroxide, barium hydroxide, calcium oxide, potassium carbonate, and the like. Sodium hydroxide, barium hydroxide and potassium hydroxide are commonly used.

Although the molar ratio of phenol to formaldehyde is critical, the other condensation reaction parameters such as time, temperature, pressure, catalyst concentrations, reactant concentrations, etc., are not critical. These parameters can be adjusted to obtain a phenolic resol of desired molecular weight and dispersibility.

The reaction of phenol and formaldehyde is generally carried out at temperatures ranging from about 70-90°C.

Resol synthesis is not pressure critical and can be carried out at reduced, elevated or normal pressure.

The catalyst concentration recommended by the authors of the study is in the range of 0.01-0.020 mol of base per mol of phenol.

The condensation reaction time can vary depending on the temperature, concentration of reactants and the amount of catalyst used. Typically, reaction times range from 8 to 20 hours.

Since the condensation reactions are catalyzed with a base, the resulting resol has an alkaline pH value. Therefore, it is desirable to be able to adjust the pH of the phenolic resol to a value of about 4.5-7.0 in order to inhibit further condensation reactions in time. The pH is adjusted by adding an acid. Hydrochloric acid, sulfuric acid, phosphoric acid, acetic acid, oxalic acid and formic acid are used for pH adjustment.

Any suitable foaming agent can be used. It should be remembered that the thermal conductivity of the resulting foam is directly related to the thermal conductivity of the blowing agent. Therefore, although substances such as n-pentane, methylene chloride, chloroform and tetrachlorocarbon can be used as blowing agents, they are not preferred, as there are fluorocarbons which

significantly better. In addition, fluorocarbon blowing agents are not soluble in phenolic foam and therefore will not diffuse over time, while some of the aforementioned blowing agents have some solubility in phenolic foam and therefore may diffuse over time. The authors of the study recommend the following as blowing agents: dichlorodifluoromethane, dichlorotetrafluoroethane; 1,1,1-trichloro-2,2,2,2-trifluoroethane; trichloromonofluoromethane; and trichlorotrifluoroethane. Carbon halides having boiling points from 5 to 55° C at a pressure of 760 mmHg are used. At the same time from 5 to 20% of the mass of the finished phenolic foam is the mass of Freon trapped in the pores.

In a complex multicomponent system consisting of substances of completely different polarity and solubility, a great deal of attention has been paid to surfactants.

In the preparation of phenolic foams, the surfactant must have properties that allow it to efficiently create an emulsion of aqueous resol, blowing agent, catalyst and additional additives to the foam composition. To prepare a good foam, the surfactant must reduce surface tension and stabilize the cells in the foamed regions prior to curing.

The authors found that nonionic, non-hydrolyzable silicone surfactants do an excellent job. Specific examples of suitable silicone surfactants include L-7003, L-5350, L-5420, L-5340 from Union Carbide Corporation and SF-1188 silicone surfactant from General Electric Company.

Typically, the amount of surfactant is in the range of from about 0.1 to about 10% by weight of the foamed resin composition. Typically, the amount of surfactant is in the range of from about 1 to about 6% by weight of the composition. The surfactant is individually mixed with the resol, foaming agent, catalyst. or foaming agent before mixing with the other components. Preferably, a portion of the surfactant is premixed with the fluorocarbon foaming agent premixed with the resol.

Although water is believed to be the main cause of cell wall degradation, the presence of water in resols is unavoidable. The authors believe that, first of all, it is very difficult and expensive to produce a resol that would have very little water. It should not be forgotten that water is formed during the reaction. In addition, water can give the resols the necessary viscosity and dispersibility. Anhydrous resols are very viscous and difficult to foam. In addition, without water, under convection heating conditions, it is difficult to control the heat release during the reaction. Accordingly, water is necessary in the composition of the foamed resolol. The amount of water present in the phenolic resol used for foaming should be 7-16% by weight of the foaming composition. If the foaming composition contains an uneven distribution of water, it will cause the walls of the phenolic resin foam to collapse and thus reduce the thermal insulation properties.

The acid component of the catalyst catalyst of the resin-based foaming composition may be any strong organic or inorganic acid having a pKa of less than about 2.0. Examples of strong inorganic acids include hydrochloric acid, sulfuric acid, phosphoric acid, nitric acid; examples of strong organic acids include trichloroacetic acid, picric acid, benzene sulfonic acid, toluene sulfonic acid, xylene sulfonic acid, phenol sulfonic acid, methanesulfonic acid, ethanesulfonic acid, and the like.

In addition to the components necessary to produce phenolic foam, the foaming mixture may be supplemented with components modifying mechanical properties, such as plasticizers, e.g., phthalic acid based in an amount of from 0.5 to 5%, resorcinol or urea to bind excess formaldehyde, flame retardants, chipping modifiers, and the like.

1.3 Conclusions from Chapter 1.

1.   Phenol formaldehyde foams have unique performance properties.

2.   The existing technologies of phenol-formaldehyde foam plastics production are inefficient and have approximately the same disadvantages: multistage, labor-intensive, time-consuming, low atomic efficiency due to the use of foaming agents, modifiers, surfactants, catalysts, which far from always meet modern requirements of environmental safety.

3.   Application of modern microwave technologies can solve a number of technological problems and make the production of phenol-formaldehyde foams cost-effective.

2 PRODUCTION OF PHENOL-FORMALDEHYDE FOAMS AND MATERIALS BASED ON THEM UNDER MICROWAVE ACTIVATION CONDITIONS

2.1 Influence of synthesis conditions on the structure of the obtained materials

The synthesis of phenol-formaldehyde foams was carried out in a microwave oven from two components: crystalline phenol and urotropine. No further reagents were used.

The method of synthesis of phenol-formaldehyde resins of resolic and novolac type under microwave irradiation conditions has been developed and is known. To obtain porous phenol-aldehyde polymers it was decided to use dry urotropine instead of aqueous formaldehyde. As is known, urotropine (hexamethylenetetramine) is a condensation product of ammonia and formaldehyde and is a colorless and odorless white substance, non-toxic, used in medicine for the treatment of urogenital infections and known as a dry propellant. When heated, urotropine dissociates into ammonia and formaldehyde. Both products are gases. Moreover, formaldehyde is the reactant and ammonia, as we assume, plays the role of the main catalyst for phenol-formaldehyde condensation.

It should be emphasized that, although the phenolurotropin mixture has been used to produce phenol-formaldehyde resins, it has never been used to produce porous polymers. This is our know-how.

To carry out this process, mixtures were prepared in which the weight ratio of phenol and urotropine was varied sequentially 9:1; 8:2; 7:3; 6:4; 5:5; 4:6; 3:7; 2:8; 1:9. The ratio of reagents influenced a number of factors: pore size, density, hardness, and brittleness.

In our opinion, the material obtained by interaction of 8 weight parts of phenol and 2 parts of urotropine was the most successful.

The reaction was carried out at all microwave power values (100, 200, 350, 500, 600, 800 W). It was found that the use of microwave irradiation with power less than 200

W is not effective, and more than 500 W leads to a number of side processes. The most successful powers were 200, 350, 500 W.

The time, depending on the power varied from 1 to 5 minutes. The reaction carried out by us under microwave activation conditions is approximately 250 times faster than the best classical samples.

For carrying out the reaction it was necessary to choose the dishes. It is not desirable to carry out the reaction in glass vessels, because it is impossible to extract the obtained polymer without damaging it, because it adheres very strongly to the walls of the vessel. 26

This also applies to paper cups. Teflon cups turned out to be the most convenient utensils for receiving. From them the ready polymer is removed practically without damage. Silicone dishes can also be used. Polyethylene and polypropylene vessels turned out to be insufficiently heat-resistant.

Each experiment was repeated several times. The results of the study are well reproducible.

Figure 2.1 - Sample No. 1. Obtained by interaction of 8 parts of phenol and 2 parts of urotropine. Power 200 W, time 3 minutes.

Figure 2.2 - Sample No. 2. Obtained by interaction of 8 parts of phenol and 2 parts of urotropine. Power 200 W, time 2.5 minutes.

Comparison of sample 1 and sample 2 clearly shows how much 30 seconds means in microwave-activated synthesis. The sample synthesized in 2.5 minutes is three times smaller than the sample synthesized in 3 minutes.

Figure 2.3 - Sample #3. Obtained by interaction of 8 parts of phenol and 2 parts of urotropine. Power 200 W, time 3.5 minutes.

Irradiation of sample 3 for 30 seconds longer than sample 1 leads to a significant overheating of the reaction mass, which leads to the complete consumption of urotropine, which leads to the cessation of the release of gases that maintain the shape of the sample. As a consequence, the overheated reaction mass, not having time to solidify, settles.

Comparing samples 4 and 5, it can be concluded that sample 4, on the contrary, was too weakly heated.

Figure 2.4 - Sample #4. Obtained by interaction of 9 parts of phenol and 1 part of urotropine. Power 200 W, time 2 minutes.

Increasing the power to 300 W allowed the product to be produced in the desired size. The maximum size was determined by the dimensions of the mold in which they were made.

Figure 2.5 - Sample No. 5. Obtained by interaction of 9 parts of phenol and 1 part of urotropine. Power 500 W, time 2 minutes.

Figure 6 shows a sample synthesized from 7 parts phenol and 3 parts urotropine at 500 W power.

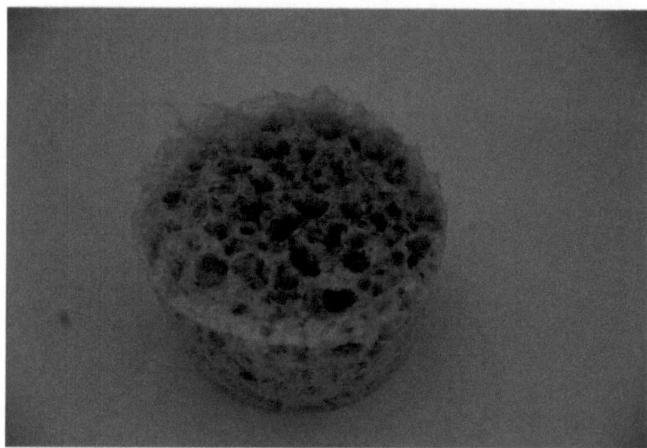

Figure 2.6 - Sample No. 6. Obtained by interaction of 7 parts of phenol and 3 parts of urotropine. Power 500 W, time 2 minutes.

Increasing the exposure time to 3 minutes and irradiation power of 500 W, allows to obtain a sufficiently voluminous sample at the ratio of phenol and urotropine parts 6:4.

Figure 2.7 - Sample No. 7. Obtained by interaction of 6 parts of phenol and 4 parts of urotropine. Power 500 W, time 3 minutes.

Increasing the mass fraction of urotropin above 50% has a negative effect on sample volumes even when the exposure is increased to 4 minutes and 500 W power.

Figure 2.8 - Sample No. 8. Obtained by interaction of 5 parts of phenol and 5 parts of urotropine. Power 500 W, time 4 minutes.

The product obtained by interaction of 9 parts of urotropine and 1 part of phenol consists mostly of unreacted urotropine and is easily destroyed even by slight mechanical impact. All other products are objects of varying degrees of hardness.

Figures 2.1-2.8 show that the most voluminous samples are 1,5,6,7. Sample 1 containing 8 mass parts of phenol turned out to be the strongest.

Samples with the content of urotropine in the initial mixture more than 50% contained unreacted urotropine in their composition. This was established by microphotography of the synthesized objects.

The structure of the synthesized samples was studied using both a Nikon 3100 camera with magnifying attachments and an Altami Bio-8 binocular microscope, magnification 40 times. The microscopic study allowed not only to reveal inclusions of unreacted urotropine, but also to obtain photographs of the surface of successful samples. The material has developed longitudinal and transverse bonds, which give rigidity to the synthesized sample.

As described in the first part of this report, the bubbles formed containing a particular thermal insulating material play an important role in the thermal conductivity. In micrographs of the resulting material, it is also possible to observe the bubbles formed, which will provide the thermal insulation properties of the material. In some cases, burst bubbles can be observed. Analysis of the synthesis conditions of the obtained substances will help to understand the necessary conditions for their production.

Figure 2.9 shows a photograph of sample 9 (4 parts phenol, 6 parts urotropine). The photos clearly show white crystals, which are unreacted urotropine (magnification 5 times).

Figure 2.9 - Sample 9. Obtained by the interaction of 4 parts phenol and 6 parts urotropine. Power 500 W, time 4 minutes.

Figure 2.10 shows photos of sample No. 2 (magnification 10 and 15 times). The photo shows a solidified surface with developed longitudinal-transverse bonds. It can be seen that most of the cells of the upper layer are perforated, while the cells of the inner layers are mostly intact. This means that the structure of this sample allows us to expect thermal insulation properties. Photos were taken with Altami Bio-8 microscope

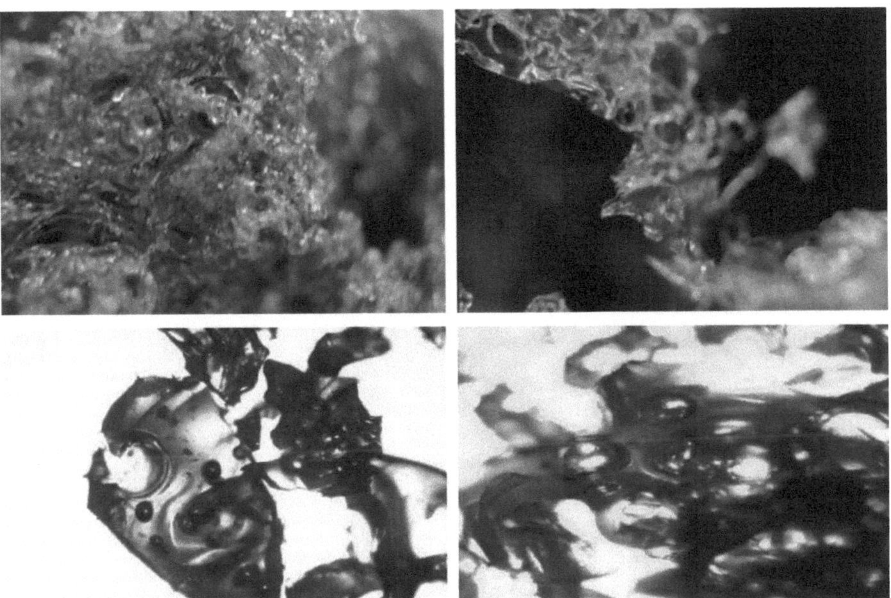

Figure 2.10 - Microphotographs of samples #2 (40x magnification).

2.2   Performance testing

Performance tests were carried out for most of the samples. During the tests it was found that the obtained experimental samples of phenol-formaldehyde foam polymer have good performance characteristics comparable to the industrially used foam plastic. Thus, the thermal conductivity coefficient of sample #2 is 0.064 Вт/(K·м), while for the reference foam this value is 0.04 Вт/(K·м). The obtained samples have thermal conductivities ranging from 0.064 to 0.095 W/(K m). These values are slightly worse than the foam, but we are confident that we will be able to reduce the pore size and approach the foam in terms of thermal conductivity. The mechanical strength of sample No. 2 is 0.64MPa. At the same time, the obtained material is superior to foam plastic in hardness. The mechanical strength of the samples could be changed in a very wide range from 0.5 to 4 kg/cm^2 , changing the ratio of reagents and synthesis conditions.

At the same time, the obtained polymer, unlike foam plastic, does not burn. It can be used to replace foam plastic in construction, where fire resistance is required. For example, when insulating premises. Figure 11 shows photos of the burned polymer. Since it does not burn itself, it was doused with automobile gasoline and set on fire. As the gasoline burned off, the polymer charred and self-extinguished. The project will continue to vary the performance properties by adding various additives.

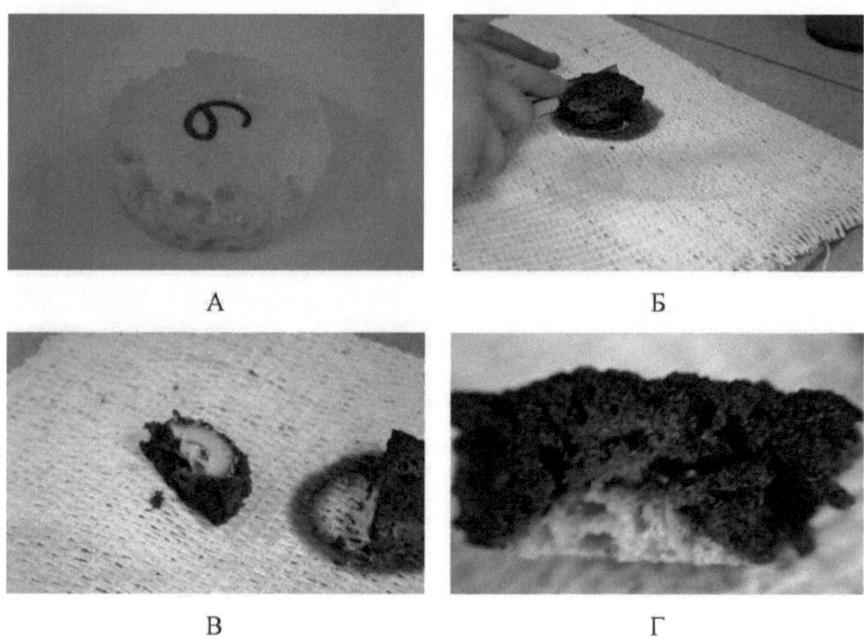

Figure 2.11 - Change of polymer after burning with gasoline. Initial polymer (a), polymer after gasoline burning (b), polymer cut into two parts (c), polymer close-up (d).

2.3 Search for optimal conditions for foam production[1]

The properties of the obtained samples depend on the composition of the initial mixture and manufacturing conditions. Sample No. 2 showed the best performance properties, so it was decided to find the optimal synthesis conditions for it. For this purpose, the orthogonal central compositional planning of the experiment for a two-factor system was used. Based on this method, the planning of the experiment was carried out. The calculated data are given in tables 2.3-2.5. Calculations, graphing

---

1 The authors are grateful to Vladimir Fedorchenko and Anastasia Khrustaleva for mathematical modeling

were carried out using Microsoft Excel 2013 program.

Calculation of the planning matrix

$$b_0^* = 88{,}33 + 80{,}67 + 89{,}67 + 83{,}00 + 82{,}67 + 93{,}67 + 90{,}67 + 85{,}67 + 88{,}67)/9 = 87{,}00$$

$$b_1 = ((-1)\times 88{,}33 + 80{,}67 + (-1)\times 89{,}67 + 83{,}00 + 82{,}67 + (-1)\times 93{,}67)/6 = -4{,}22$$

$$b_2 = ((-1)\times 88{,}33 + (-1)\times 80{,}67 + 89{,}67 + 83{,}00 + 90{,}67 + (-1)\times 85{,}67)/6 = 1{,}44$$

$$b_{12} = (88{,}33 + (-1)\times 80{,}67 + (-1)\times 89{,}67 + 83{,}00)/4 = 0{,}25$$

$$b_{11} = ((0{,}33)\times(88{,}33 + 80{,}67 + 89{,}67 + 83{,}00 + 82{,}67 + 93{,}67) -$$

$$-(0{,}67)\times(90{,}67 + 85{,}67 + 88{,}67))/2 = -2{,}0$$

$$b_{22} = ((0{,}33)\times(88{,}33 + 80{,}67 + 89{,}67 + 83{,}00 + 90{,}67 + 85{,}67+) - (0{,}67)\times$$

$$\times(82{,}67 + 93{,}67 + 88{,}67))/2 = -2{,}0$$

$$b_0 = b_0^* - b_{11}\frac{6}{9} - b_{22}\frac{6}{9} = 87{,}00 - (-2{,}0)\times 2/3 - (-2{,}0)\times 2/3 = 89{,}67$$

Let's calculate y_{pac4et} , for this purpose let's substitute the previously obtained coefficients into the obtained regression equation:

$$y_1 = 89{,}67 - 4{,}22\times(-1) + 1{,}44\times(-1) + 0{,}25\times 1 - 2{,}0\times 1 - 2{,}0\times 1 = 88{,}33$$

$$y_2 = 89{,}67 - 4{,}22\times 1 + 1{,}44\times(-1) + 0{,}25\times(-1) - 2{,}0\times 1 - 2{,}0\times 1 = 80{,}6$$

Table 2.1 - Extended planning matrix for 27 experiments in Excel program

№	x1, % urotropin		x2, t		x1*	x2*	X1X2	Y	Y race.	Y medium
	code	significance	code	significance						
1	2	3	4	5	6	7	8	9	10	11
1	-1	20,00	-1	2,50	0,33	0,33	1	88	88,69	
2	-1	20,00	-1	2,50	0,33	0,33	1	87	88,69	88,33
3	-1	20,00	-1	2,50	0,33	0,33	1	90	88,69	
4	1	40,00	-1	2,50	0,33	0,33	-1	80	79,75	
1	2	3	4	5	6	7	8	9	10	80,67
5	1	40,00	-1	2,50	0,33	0,33	-1	80	79,75	
6	1	40,00	-1	2,50	0,33	0,33	-1	82	79,75	
7	-1	20,00	1	3,50	0,33	0,33	-1	90	91,08	

8	-1	20,00	1	3,50	0,33	0,33	-1	91	91,08	89,67
9	-1	20,00	1	3,50	0,33	0,33	-1	88	91,08	
10	1	40,00	1	3,50	0,33	0,33	1	85	83,14	
11	1	40,00	1	3,50	0,33	0,33	1	83	83,14	83,00
12	1	40,00	1	3,50	0,33	0,33	1	81	83,14	
13	1	40,00	0	3,00	0,33	-0,67	0	83	83,44	
14	1	40,00	0	3,00	0,33	-0,67	0	82	83,44	82,67
15	1	40,00	0	3,00	0,33	-0,67	0	83	83,44	
16	-1	20,00	0	3,00	0,33	-0,67	0	95	91,89	
17	-1	20,00	0	3,00	0,33	-0,67	0	92	91,89	93,67
18	-1	20,00	0	3,00	0,33	-0,67	0	94	91,89	
19	0	30,00	1	3,50	-0,67	0,33	0	90	89,11	
20	0	30,00	1	3,50	-0,67	0,33	0	90	89,11	90,67
21	0	30,00	1	3,50	-0,67	0,33	0	92	89,11	
22	0	30,00	-1	2,50	-0,67	0,33	0	85	86,22	
23	0	30,00	-1	2,50	-0,67	0,33	0	86	86,22	85,67
24	0	30,00	-1	2,50	-0,67	0,33	0	86	86,22	
25	0	30,00	0	3,00	-0,67	-0,67	0	89	89,67	
26	0	30,00	0	3,00	-0,67	-0,67	0	90	89,67	88,67
27	0	30,00	0	3,00	-0,67	-0,67	0	87	89,67	

$$y_3 = 89,67 - 4,22 \times (-1) + 1,44 \times 1 + 0,25 \times (-1) - 2,0 \times 1 - 2,0 \times 1 = 89,67$$

$$y_4 = 89,67 - 4,22 \times 1 + 1,44 \times 1 + 0,25 \times 1 - 2,0 \times 1 - 2,0 \times 1 = 83,00$$

$$y_5 = 89,67 - 4,22 \times 1 + 1,44 \times 0 + 0,25 \times 0 - 2,0 \times 1 - 2,0 \times 0 = 82,67$$

$$y_6 = 89,67 - 4,22 \times (-1) + 1,44 \times 0 + 0,25 \times 0 - 2,0 \times 1 - 2,0 \times 0 = 93,67$$

$$y_7 = 89,67 - 4,22 \times 0 + 1,44 \times 1 + 0,25 \times 0 - 2,0 \times 0 - 2,0 \times 1 = 90,67$$

$$y_8 = 89,67 - 4,22 \times (-1) + 1,44 \times 0 + 0,25 \times 0 - 2,0 \times (-1) - 2,0 \times 0 = 85,67$$

$$y_9 = 89,67 - 4,22 \times 0 + 1,44 \times 0 + 0,25 \times 0 - 2,0 \times 0 - 2,0 \times 0 = 88,67$$

Table 2.2 - Selection of factors and variation interval for the matrix in Excel program

Characterization	x1, % urotropin	x2, t (min)

	code	significance	code	Significance
Lower level	-1	20	-1	2,5
Basic level	0	30	0	3
Upper level	1	40	1	3,5
Variation interval		10 %		0.5 min.

Table 2.3 - Determination of regression equation coefficients for the matrix in Excel program

Determination of the coefficients of the regression equation						
b0*	Б1	Б2	Б12	Б11	Б22	Б0
87,00	-4,22	1,44	0,25	-2,00	-2,00	89,67

Let's consider private dependences, which will allow to receive graphical dependences of influence of investigated factors on efficiency of substance output.

Consider the dependence $Y = f(X_1)$, for fixed values of $X_2 = -1; 0; +1$.

at $X_2 = -1$

$$y = 89,67 - 4,22 \times X_1 + 1,44 \times X_2 + 0,25 \times X_1 X_2 - 2 \times X_1^2 - 2 \times X_2^2 =$$

$$= 89,67 - 4,22 \times X_1 + 1,44 \times (-1) + 0,25 \times X_1 \times (-1) - 2 \times X_1^2 - 2 \times (-1)^2 =$$

$$= 86,23 - 4,47 \times X_1 - 2 \times X_1^2$$

Let's pass to the real values of the process, for this purpose let's substitute by the formula

$$X_1 = (a - x_{10})/\Delta,$$

where X_1 - coded value of the factor; a -% ratio of urotropine to phenol, X_{10} - factor value in the center of the plan, %; Δ - variation interval, %.

$$X_1 = \frac{a - 30}{10}$$

$$X_1 = 0{,}1 \times a - 3$$

$$y=86{,}23 + 4{,}47\times(0{,}1\times a{-}3) - 2\times(0{,}1\times a{-}3)^2 =86{,}23+0{,}447\times a+13{,}41 - 2\times((0{,}1\times a)^2 - 2\times0{,}1\times a\times3+9)=99{,}64-0{,}447\times a+0{,}02 \times a^2+1{,}2\times a-18$$

$$y=81{,}64+0{,}753\times a - 0{,}02\times a^2$$

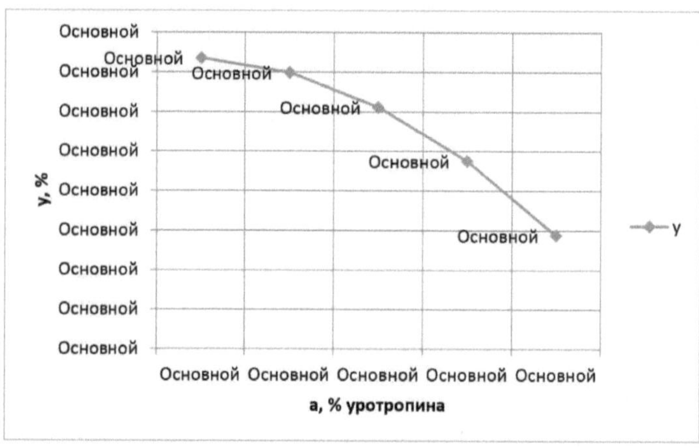

Figure 2.12 - Plot of partial dependence of substance yield on % ratio of urotropine to phenol $y(x_1)$ at $x_2 = -1$

at $X_2 = 0$

$$y=89{,}67- 4{,}22\times X_1+1{,}44\times(0) +0{,}25\times X_1\times(0) - 2\times X_1^2 -2\times(0)^2 =$$

$$= 86{,}23 - 4{,}47\times X_1 - 2\times X_1^2$$

Expression in real values: $\qquad$ $y=89.67 - 4.22\times(0.1\,ha{-}3) - 2\times(0.1\,ha{-}3)^2 =$

$=102.33-0.422\times a-0.02\times a^2 +1.2\times a-18u=84 \quad .33+0.778\times a^2$

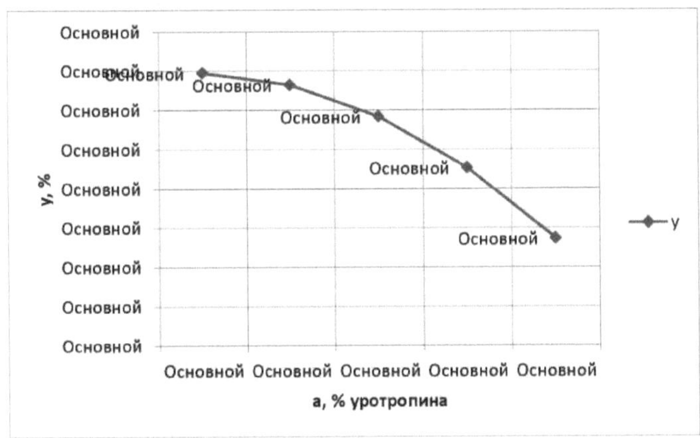

Figure 2.13 - Plot of partial dependence of substance yield on % ratio of urotropine to phenol $y(x_1)$ at $x_2 = 0$

when X2 = +1

$$y = 89,67 - 4,22 \times X_1 + 1,44 \times (+1) + 0,25 \times X_1 \times (+1) - 2 \times X_1^2 - 2 \times (+1)^2 =$$

$$= 89,11 - 3,97 \times X_1 - 2 \times X_1^2$$

Expression in real values:

$$y = 89,11 - 3,97 \times (0,1 \times a - 3) - 2 \times (0,1 \times a - 3)^2 = 101,02 - 0,397 \times a - 0,02 \times a^2 + 1,2 \times a - 18$$

$$y = 83,02 + 0,803 \times a - 0,02 \times a^2$$

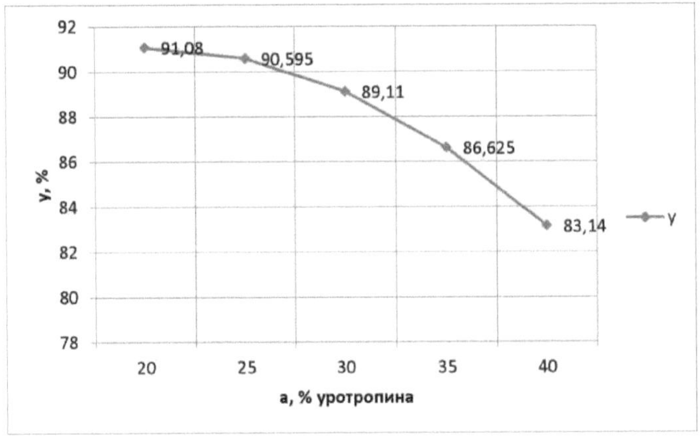

Figure 2.14 - Plot of partial dependence of substance yield on % ratio of urotropine to phenol $y(x_1)$ at $x_2 = +1$

The partial dependences of the substance yield on the emission time are obtained at fixed values of the factor x_1, equal to -1, 0, +1 We obtain three partial regression equations of the form:

$X_1 = -1$

$$y = 89{,}67 - 4{,}22 \times X_1 + 1{,}44 \times X_2 + 0{,}25 \times X_1 X_2 - 2 \times X_1^2 - 2 \times X_2^2 =$$
$$= 89{,}67 - 4{,}22 \times (-1) + 1{,}44 \times X_2 + 0{,}25 \times (-1) \times X_2 - 2 \times (-1)^2 - 2 \times X_2^2 =$$
$$= 91{,}89 + 1{,}19 \times X_2 - 2 \times X_2^2$$

Let's pass from coded to real values of emission time, for this purpose we will use the expression

$$X_2 = (t - x_{20})/\Delta$$

X_2 - coded value of the factor; t - radiation time, min; X_{20} - factor value in the center of the plan, min; Δ - variation interval, min.

$$X_2 = \frac{t - 3}{0{,}5}$$
$$X_2 = 2 \times t - 6$$

$$y = 91{,}89 + 1{,}19 \times (2 \times t - 6) - 2 \times (2 \times t - 6)^2 = 91{,}89 + 2{,}38 \times t - 7{,}14 - 2 \times ((2 \times t)^2 - 2 \times 2 \times t$$
$$\times 6 + 36) = 91{,}89 + 2{,}38 \times t - 7{,}14 - 2 \times t^2 + 12 \times t - 18$$
$$y = 66{,}75 + 14{,}38 \times t - 2 \times t^2$$

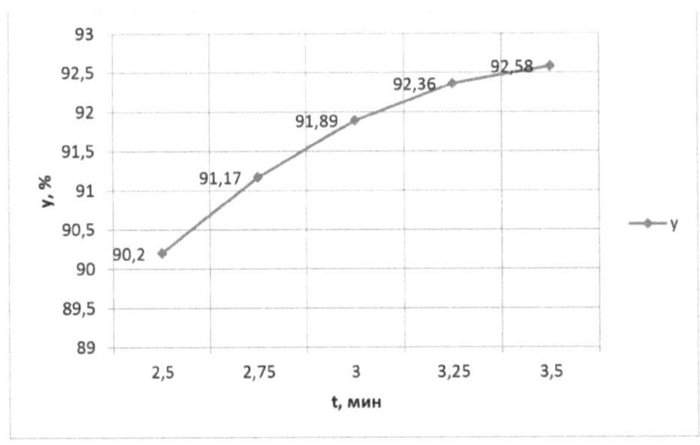

Figure 2.15 - Plot of partial dependence of substance yield on irradiation time $y(x_2)$ at $x_1 = -1$

$$X_1 = 0$$

$$y=89{,}67 - 4{,}22\times(0) +1{,}44\times X_2+0{,}25\times(0)\times X_2 - 2\times(0)^2 - 2\times X_2^2 =$$
$$= 89{,}67+1{,}44\times X_2 - 2\times X_2^2$$

Transition to real values of emission time:

$$y=89{,}67 +1{,}44\times(2\times t-6)- 2\times(2\times t-6)^2= 89{,}67+2{,}88\times t-8{,}64- 2\times t^2+12\times t-18$$
$$y=63{,}03+14{,}88\times t-2\times t^2$$

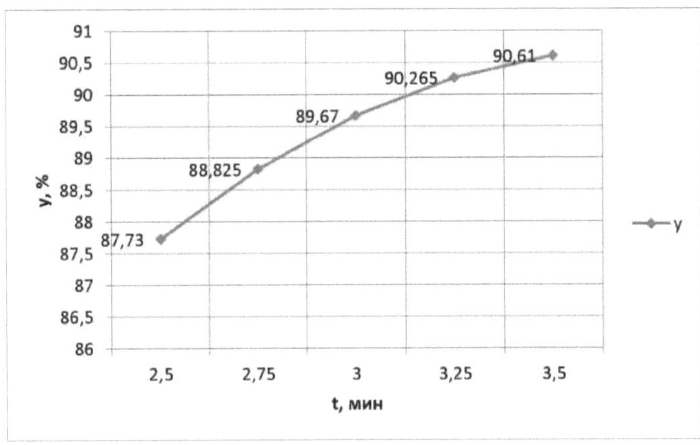

Figure 2.16 - Plot of partial dependence of substance yield on irradiation time $y(x_2)$ at $x_1 = 0$

$X_1 = +1$

$$y = 89,67 - 4,22 \times (+1) + 1,44 \times X_2 + 0,25 \times (+1) \times X_2 - 2 \times (+1)^2 - 2 \times X_2^2 =$$

$$= 83,45 + 1,69 \times X_2 - 2 \times X_2^2$$

Transition to real values of emission time:

$$y = 83,45 + 1,69 \times (2 \times t - 6) - 2 \times (2 \times t - 6)^2 = 83,45 + 3,38 \times t - 10,14 - 2 \times t^2 + 12 \times t - 18$$

$$y = 55,31 + 15,38 \times t - 2 \times t^2$$

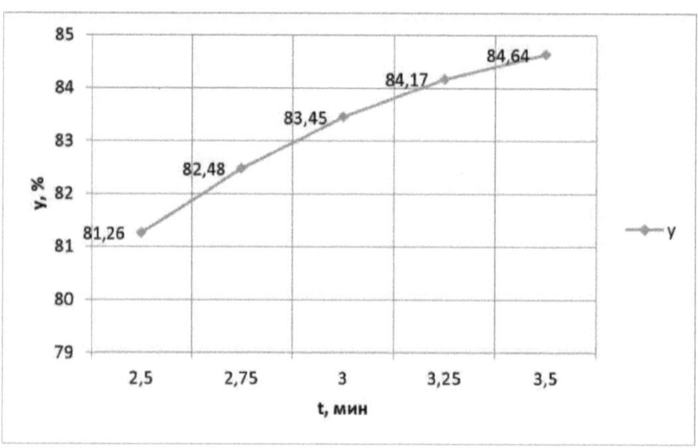

Figure 2.17 - Plot of partial dependence of substance yield on irradiation time $y(x_2)$ at $x_1 = +1$

Finding the optimum.

Solving the system of equations, we obtain the values of x_1 and x_2 (in coded form

$$\left.\begin{cases} \dfrac{dy}{dx_1} = 0 + b_1 + b_{12}x_2 + 2b_{11}x_1 = 0 \\[2mm] \dfrac{dy}{dx_2} = b_2 + b_{12}x_1 + 2b_{22}x_2 = 0 \end{cases}\right\} =$$

$$= \begin{cases} -4,22 + 0,25x_2 - 4x_1 = 0 \\ 1,44 + 0,25x_1 - 4x_2 = 0 \end{cases}$$

$$x_1 = -1,037$$

$$x_2 = 0,288$$

Let's move from coded values to real values

$$X_1 = 0,1 \times a - 3$$

$$-1,037 = 0,1 \times a - 3$$

$$a = 19,63 \%$$

$$X_2 = 2 \times t - 6$$
$$0,288 = 2 \times t - 6$$

$$t = 3.144 \text{ min.}$$

Substituting the values $X_1 = -1.037$ and $X_2 = 0.288$ into the equation, we get the value of product yield

$$y = 89,67 - 4,22 \times X_1 + 1,44 \times X_2 + 0,25 \times X_1 X_2 - 2 \times X_1^2 - 2 \times X_2^2 =$$
$$= 89,67 - 4,22 \times (-1,037) + 1,44 \times 0,288 + 0,25 \times (-1,037) \times 0,288 - 2 \times (-1,037)^2 -$$
$$2 \times (0,288)^2 = 92,071\%$$

2.4 Conducting initial environmental safety tests

For the analysis, a 10 g sample was crushed using a mechanical porcelain mortar and pestle and mixed in a beaker with distilled water for one hour at room temperature. The resulting solution was analyzed according to international methods for the presence of phenol (ISO 6439:1990), ammonia (ISO 5664:1984) and formaldehyde (ISO 14184-1:1999).

The phenol content ranged from 0.0003 to 0.0009 mg/l; ammonia from 0.1 to 0.5 mg/l; formaldehyde from 0.01 to 0.09. Thus, the phenol content in all samples slightly (3-9 times) exceeds the MPC; ammonia concentration does not reach the MPC; formaldehyde concentration, depending on the sample in some cases slightly exceeds the norm.

The literature review showed that it is normal practice to have free phenol and formaldehyde in freshly prepared phenol-formaldehyde foams from 3 to 7% of their mass [17]. Of course, such a high content of toxic substances in phenol-formaldehyde foams presented a serious problem, the successful solution of which has been described in a series of patents [18-19]. Removal of free formaldehyde and phenol by heating the obtained polymer under vacuum at a temperature of about $150°$ C for 4-6 hours is considered to be the most effective. In this case the polymer with phenol and formaldehyde content below MPC is obtained

2.5 Preparation of polymers containing an external skeleton

Due to the development of green chemistry, as well as green technologies in construction, sandwich panels, which are a very cheap and popular building material, have become very popular. The most popular technology of making sandwich panels is to glue a hard shell (wood or plastic) on the heat-insulating material. For example, on foam plastic. So make plastic doors, panels for insulation of balconies, often from such material are made street kiosks. It should be remembered that the most commonly used material - Styrofoam is flammable and extremely unstable to organic solvents.

We tested a fundamentally different method of obtaining heat-insulating material. For this purpose, plates were cut out of plastic, from which a box with dimensions of 10x10x2.5 cm was made. Holes were made in the lid to allow the removal of gases. A mixture of phenol and urotropine was poured into the box and it was placed in a microwave oven. The irradiation was carried out at a power of 500 W and a time of 4 minutes. The reaction produced a porous phenol-formaldehyde that rigidly glued all parts of the plastic to form a single fragment. This experiment showed that in this way can be made similar products from dielectric materials transparent to microwave radiation (wood, ceramics, plywood) with adhesion to resin. Also, this technology can be used to produce products of complex shapes, which is impossible with the current technologies. The stages of production are illustrated in Figures 2.18-2.20.

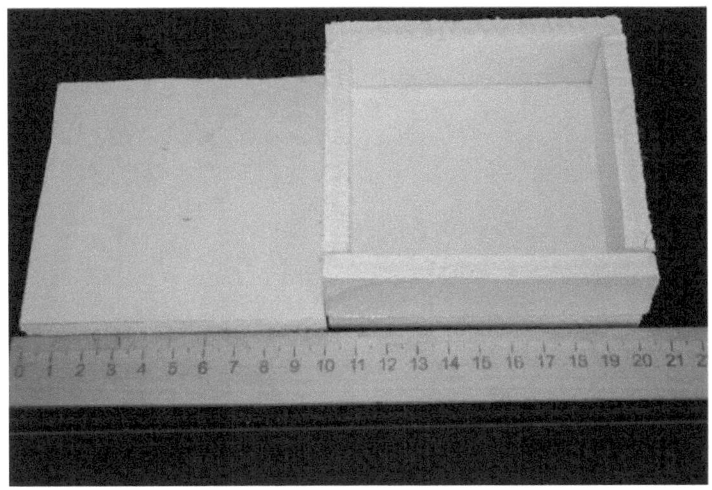

Figure 2.18 - Frame made of polymer material

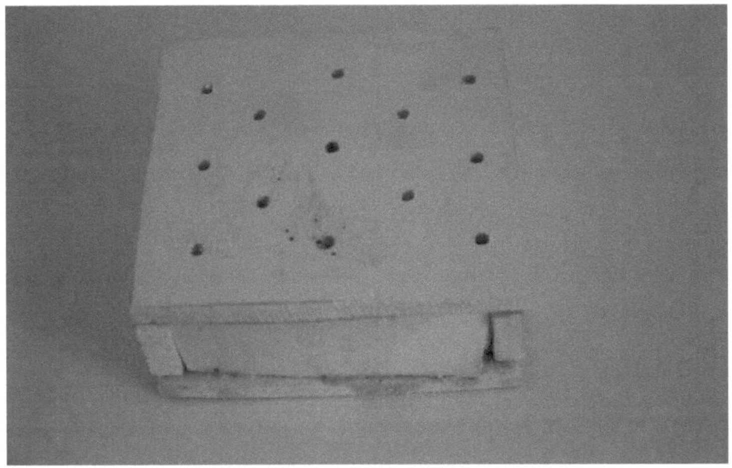

Figure 2.19 - Frame after the reaction has been carried out

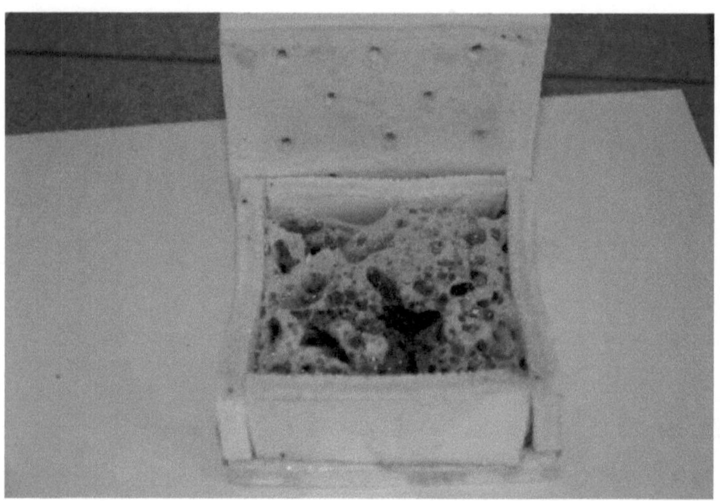

Figure 2.20 - The frame with the resulting phenol-formaldehyde resin-type foam. The top lid is firmly glued to the box during the reaction.

2.6 Obtaining "laminated" polymers.

Porous phenol-formaldehyde polymers are characterized by brittleness on one side and an uneven surface on the other. The performance of the material can be improved by finding a way to make the surface smooth and even.

Two methods were used to produce laminated polymers. The first, mechanical, consisted in gluing a smooth sheet of polyethylene terephthalate onto the surface of porous phenolphloromaldehyde resin. Two methods were tried.

The first one consisted in gluing, cut to size, a sheet of polyethylene terephthalate onto an uncooled polymer synthesized in a rectangular shape. Upon cooling, the porous phenol-formaldehyde resin solidifies to form a material with a smooth polymer surface. A sheet of plywood, kragis, ceramic tiles, and glass have been glued in this way.

The second method consisted of bonding sheets of plywood, kragis, ceramic tiles or glass to a cooled rectangular mass of porous phenol-formaldehyde polymer using available adhesives such as Titan Fix.

The second, more interesting from the chemical point of view, method of obtaining laminated polymer was also tested. For this purpose, an original technology for the production of polyesters based on phthalic anhydride and polyatomic alcohols under microwave activation conditions was developed. The usual method of polyester resin production based on phthalic anhydride and ethylene glycol (or glycerol) takes 5-8 hours. The original technology developed by us takes no more than 10 minutes. And a patent for the method of producing polyester resins in MVA is currently being prepared.

Laminating consisted of applying molten polyester resin to the surface of a porous phenol-formaldehyde resin. In this process, the two polymers are very well bonded, forming a very strong bond.

The advantages of this method include high durability of the coating, fast and simple method of manufacturing polyester resin. The disadvantages include high brittleness of the polyester resin. However, additives are known that can give polyester resins flexibility. We are currently working on their production under MVA conditions.

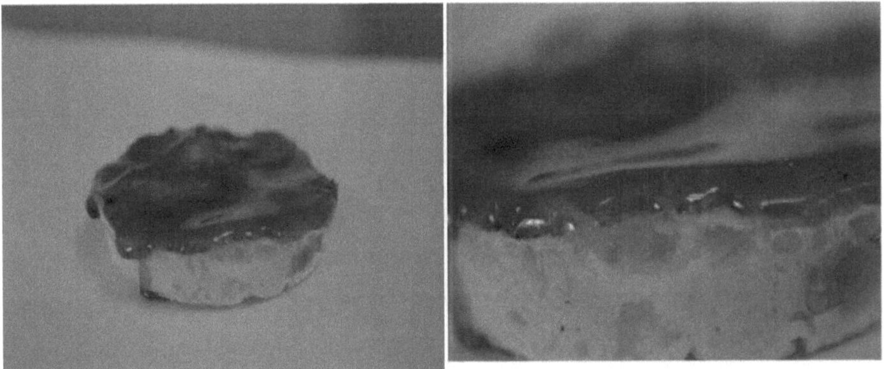

Figure 2.21 - Polyester resin layer on the surface of porous phenol-formaldehyde resin

2.7 Preparation of porous polymers with different fillers

This part of the study had two objectives: to investigate the possibility of obtaining porous phenol-formaldehyde polymers containing different fillers and to study their performance properties.

During the research, attempts were made to obtain porous polymers containing sawdust, wood chips, sand, fine stone, asbestos crumbs and asbestos cloth, organic wool, mineral wool, and unusable grain as filler.

During the study it was found that sawdust and wood chips can serve as fillers for phenol-formaldehyde resins (obtained under MVA conditions), but when obtaining porous phenol-formaldehyde resins, the resulting mass is heterogeneous: porous polymer separately, formed by the bottom layer, sawdust or chips separately, formed by the top layer. The resulting mass does not have mechanical stability and crumbles when trying to remove it from the mold.

Phenol-formaldehyde resin, formed from phenol and aqueous formaldehyde solution, does not emit gas when heated, as it happens during polymerization of phenol-urotropin mixture. Phenol-formaldehyde mass, during polymerization under MVA conditions slowly thickens and sawdust is added to it after the reaction is carried out for 2-3 minutes. Sawdust is impregnated with the resulting resin, which when cooled solidifies into a homogeneous mass.

The phenol-urotropin mixture also forms a homogeneous liquid mass when melted. However, its density is such that sawdust and chips remain on the surface of this mass. Subsequent heating is accompanied by abundant release of gas, which does not contribute to the homogenization of the mixture. As a result: sawdust separately, resin separately. The resin is very loose and easily destroyed.

The poor filler turned out to be organic wool. The obtained materials had brittleness less than the original, porous phenolaldehyde resin.

Sand or small stones (up to 1 cm in diameter) also do not form a single mass with porous phenolaldehyde resin, they settle to the bottom. This is understandable: at the stage of phenol and urotropine melting, sand and pebbles settle to the bottom and do not participate in the process of porous phenol-formaldehyde resin production.

Asbestos grit has no effect on the mechanical strength of the resulting polymer. Asbestos is impregnated with phenol-formaldehyde resin and loses its flexibility.

After curing, it is a very brittle material.

Surprisingly, good strong tiles are obtained with porous phenol-formaldehyde resin if substandard grains are used as filler. As you can see in the photo, the grains are tightly pressed together. Bonding occurs throughout the entire volume, and the resulting mass is characterized by hardness.

The microphotograph taken by Altami microscope shows that the grains are well coated with phenol formaldehyde resin. It can also be seen that the grains are well glued together by the cured phenol formaldehyde resin.

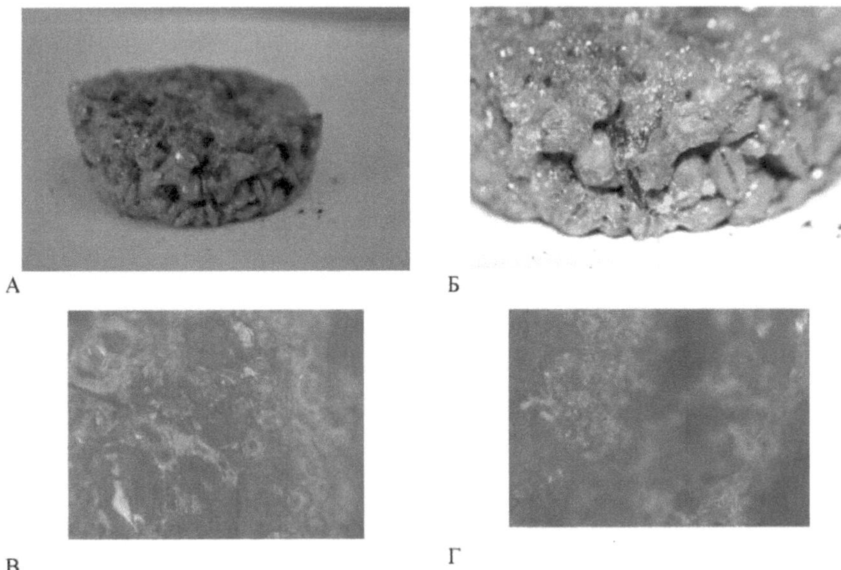

Figure 2.22 - Porous phenol-formaldehyde resin with filler from substandard grains (A - general view of the material; B - life-size macrophotograph; C - grain covered with resin (30 times magnification); D - resin connecting grains).

Mineral wool (Ursa) has shown the best results as a filler. In normal condition it is a flexible, soft, voluminous thermal insulation material consisting of mineral threads. After impregnation with phenol-aldehyde resin, this material loses volume, the released gas does not return the original volume of Ursa, but only slightly, 2-3 mm separates the layers that make up this wool. After curing, the mineral wool impregnated with phenol-formaldehyde resin turns into a very hard material. It is impossible to break or crumble this material with your hands, but it is well sawed with a metal hacksaw both along and across the fibers. This material resembles plywood in its mechanical properties.

The figures show micrographs of the polymer structure where the filler was mineral wool. It can be seen how the fully wetted mineral wool solidified. The interwoven filaments formed the outer skeleton and gave the material strength.

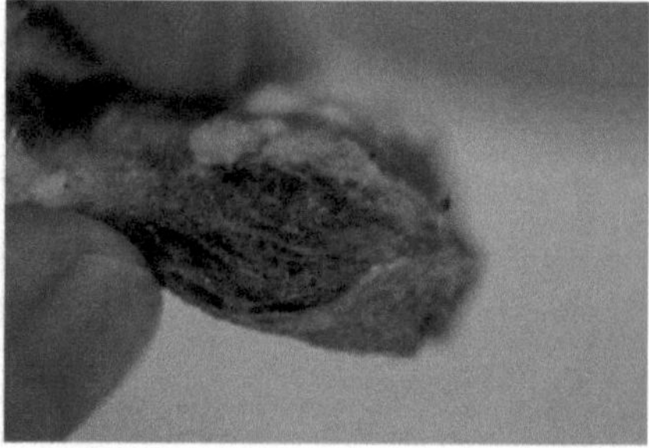

Figure 2.23 - Photographs of porous phenol-formaldehyde polymer filled with mineral wool (ursa).

The following picture shows the ursa threads with polymer beads along the length of the threads. These pictures were taken on a piece of material that has not fully polymerized.

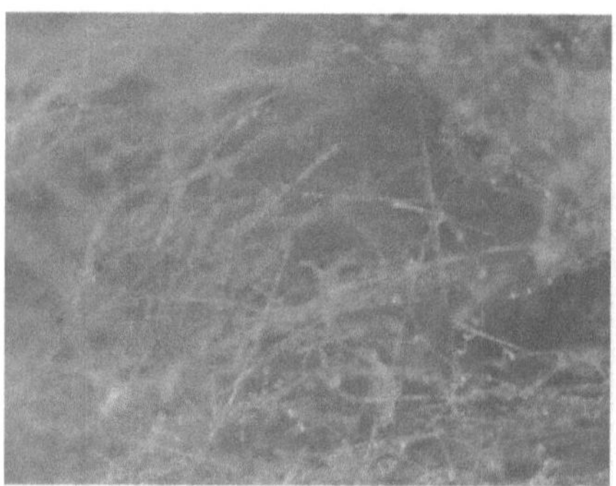

Figure 2.24 - Microphotographs of porous phenol formaldehyde polymer with mineral wool (ursa) filler.

By comparing the photographs, it is possible to speculate on the likely mechanism of formation of this material.

The first stage is impregnation and the formation of beads of molten phenol formaldehyde resin on the ursa threads.

The second stage is the joining of the beads to form a single phenol-formaldehyde coating on the surface of the filament.

The third stage is the gluing of the filaments.

Yarns bonded in this way give the material strength. While most of the obtained porous phenol-formaldehyde samples are characterized by brittleness.

3 EXPERIMENTAL PART [2]

3.1 Example of resol synthesis based on phenol and formaldehyde under convection heating conditions [11].

The molar ratio of formaldehyde to phenol 2:1 was used for the synthesis. The synthesis was carried out in a 4-liter reactor equipped with a reflux condenser, a thermocouple for temperature control, a dropping funnel, and a top-drive stirrer with double blades. The means for heating (flask heater) and cooling (ice bath) were prepared in advance. First, a solution of 1434 grams of 90% phenol (13.73 mol) was placed in the flask. Then 1207 g of powdered 91% paraformaldehyde (36.61 mol) was weighed and added to the reactor. This phenol formaldehyde mixture was stirred at 78°C heating. Meanwhile, a solution of aqueous 45% KOH was prepared. Then, 35.53 g of 45% KOH (0.285 mol) was added to 478.4 g of 90% phenol (4.58 mol) and mixed thoroughly. In this way potassium phenolate was made. The potassium phenolate mixture was loaded into a drip funnel. When the temperature in the reactor reached 78°C, the potassium phenolate solution was added dropwise for 150-minutes. During the addition of potassium phenolate, the temperature in the reactor was maintained between 78°C-80°C by heating and/or cooling the reactor. In the early stages, it is necessary to periodically cool the reactor and monitor the exothermic reaction. Also in the early stages, a small gel forms and then disappears during the addition of potassium phenolate. Close attention to temperature maintenance is necessary when a gel is formed, as heat transfer through the gel is somewhat slower.

After all of the potassium phenolate mixture was added to the reaction mixture, the reaction mixture was heated to 85°C-88°C and kept at this temperature.

At the same time every 30 minutes the viscosity is measured by Gardner. When the bubble viscosity was 60 seconds, 14.57 g of 90% formic acid was added to the solution and the reaction mixture was cooled to 55°C. After the reaction temperature reached 55°C, 190 grams of Morflex and 1129 g of dimethyl isophthalate were added to the resin and stirring was carried out until the solution was completely

---

2 The authors thank Sanakulova Saniya for her active participation in the experiments

homogenized. The reaction mixture was then transferred to a storage container and stored refrigerated until use. The resulting resol resin had a Brookfield viscosity of 6600 Santi poise at 25°C. The resol resin contained 1.9% free phenol, 3.6% free formaldehyde, and 17.3% water. The average molecular weight was 981, the average numerical molecular weight was 15507, and the dispersity value was 1.93.

3.2 Example of phenol-formaldehyde foam production [11]

Phenol formaldehyde foam is produced in the laboratory using a laboratory mold. The mold was made with one-inch aluminum bars for the sides and one-inch aluminum plates for the top and bottom lid and had internal dimensions from 4x4 inches. The dimensions can be varied. The mold is coated with release agent and preheated at 66°C in a desiccator. A piece of dry corrugated cardboard about 4x4 inches was dried in a 66°C oven for 1015 minutes. While the mold and cardboard were warmed in the oven, preparation of the resin resin for the foaming process was carried out. First, 10 parts (33.2 grams) of a 50/50 by weight mixture of Freon 11 / Freon 113 (trichloromonofluoromethane / 1,1,2,2 -trichloro-1,2,2,2-trifluoroethane) are premixed in a mixer at high stirring speed (3000 rpm) with 1 part (3.3 grams) of a silicone surfactant (Union Carbide L-7003). This mixture is placed in an ice bath and cooled to 12-15° C. At this time, 76.6 parts (254.3 grams) of the resolol obtained in the previous example were mixed under vigorous stirring with 2.4 parts (8.0 grams) of the silicone surfactant L7003. After vigorous stirring of both mixtures, they were combined and cooled in an ice bath to 10-12°C. When the desired temperature was reached, 10 parts (33.2 g) of anhydrous toluenesulfonic acid in xylene was placed in a drop funnel with an external cooling jacket and cooled to 4-5°C. Anhydrous toluenesulfonic acid, which is the catalyst in this process was mixed with previously prepared resol, foaming agent, surfactant at high speed for 10-15 seconds. Then 210 grams of the resulting resol foam composition was immediately poured into a mold. The mold was closed on all the clamps. The mold with the foaming composition was placed in an oven at 65°C for 4 minutes. After the mold is removed from the oven, the foam was removed from the mold and weighed. The foam was left for 24 hours after which it was proceeded to study its properties. The density of the obtained material was 52 kg/m^3 .

3.3 A general method for the preparation of phenol-formaldehyde foam based on phenol and urotropine under microwave activation conditions.

Pre-weighed phenol and urotropine are thoroughly rubbed in a porcelain mortar, after which these substances are placed in a silicone mold with a volume of 75-100 ml and thoroughly mixed. The obtained mixture is irradiated in a microwave oven for the required time (2-5 minutes) at a power from 200 to 500 W. After that, the obtained material is allowed to cool down and the obtained material is removed from the silicone mold.

3.4 A general method for producing phenol formaldehyde foam in an outer frame.

A box made of dielectric material (plastic, ceramic tile, plywood, wood) with the dimensions of 10x10 cm sheets and 2.5 cm thick sides was prepared in advance. The sides of the box were held together with household silicone. Holes for gas outlet were made in the top lid.

20-30 g of thoroughly crushed and mixed phenol-urotropin mixture in the ratio of 8:2 were added to the obtained container. Then the container was placed in a microwave oven for 5-7 minutes at irradiation power 200500 W. After completion of the process, foam was formed

3.5 A general method for the preparation of phenol formaldehyde foam with various fillers.

For the preparation of phenol-formaldehyde foams a mixture of phenol and urotropine in the ratio of 8:2 was used. The mass of the filler was not more than 40% of the mass of the foaming composition. Compositions with inorganic wool (ursa) and substandard grain were particularly successful.

3.6 Synthesis of glyphthalic esters under microwave activation conditions and their coating of phenol-formaldehyde foams

2.96 g of orthophthalic acid anhydride and 2.67 g of glycerol were placed in a silicone beaker. The mixture was thoroughly mixed and subjected to microwave irradiation. The exposure time and power were 10 min and 500W, respectively. After cooling slightly, the mixture was applied to the surface of phenol formaldehyde foams using a silicone brush. After curing, a hard, even, smooth surface was formed on the surface of phenol formaldehyde foams.

3.7   Definition of phenol

Phenol determinations were carried out according to ISO 6439:1990.

3.8   Ammonia determination

Ammonia determinations were performed according to ISO 5664:1984.

3.9   Determination of formaldehyde

Formaldehyde was determined according to ISO 14184-1:1999

4   CONCLUSION

1.   In the course of experiments, we have developed a methodology for the production of porous phenol-formaldehyde resin, which has no analogues in the world. The fabrication time under conditions of microwave activation of materials varied in the range from 3 to 5 minutes, which is approximately 250-400 times faster than the synthesis by known classical methods.

2.   The developed technology excludes the use of catalysts, metals, acids, alkalis, which allows to obtain high-purity polymer.

3.   The optimal conditions for the synthesis of porous phenol formaldehyde resin were found using the central orthogonal method of mathematical planning of experiment.

4.   The study established the possibility of obtaining a porous phenol-formaldehyde polymer in an external framework, which can be made of any material transparent in the microwave range.

5.   Methods for producing porous polymers with various fillers have also been developed.

6.   Analytical methods for the quantification of phenol, ammonia and formaldehyde were mastered.

7.   Methods of lamination of porous phenol-formaldehyde polymers have been developed, which makes it possible to give them a smooth durable surface. It should be noted that to solve this problem a new method of synthesizing polyesters based on phthalic anhydride was developed.

REFERENCE LIST

1.   The concept of transition of the Republic of Kazakhstan to "green economy". Astana. 30.05.2013.

2.   Galema S.A. Microwave chemistry// Chemical society Reviews.- 1997.- Vol.26.- P.233-238

3.   Mahmoud E.E., Dostlov J., Pokorn J., Lukesov D., Dolezal M.. Oxidation of Olive Oils during Microwave and Conventional Heating for Fast Food Preparation // Czech J. Food Sci.- 2009. - Vol. 27.- P. 173-177.

4.   Bogdal D. Microwave assisted Organic Synthesis: One Hundred

Reaction Procedures.- UK.:Elsevier, 2005. -P. 1-5.

5.   Berdonosov S.S. Microwave chemistry // Sorovskii

Educational Journal. - 2001. - T.7, №1. - C.32-38.

6.   Baghurst D.R., Mingos D.M.P. Superheating effects associated with microwave dielectric heating // Chem. Commun. - 1992. - P.674.

7.   Stuerga D., Gonon K., Lallemant M. Microwave heating as a new way to induce selectivity between competitive reactions. Application to isomeric ratio control in sulfonation of naphthalene // Tetrahedron. - 1993. - Vol.49. - P.62296234.

8.   Kappe C.O., Dallinger D., Murphree S.S. Practical Microwave Synthesis for Organic Chemists.-Weinheim: Wiley-VCH, 2009. -300 p.

9.   U.S. Pat. EP20070111135 USA. Loss-on-drying instrument with a microwave source and an infrared source / Collins M. J., Revesz R. N.; publ. 01/16/2008. -13c.

10.   Laszlo T.S. Industrial Applications of Microwaves// The Physics Teacher.- 1980.- P. 570-579.

11.   Khrustalev D., Gazaliev A. Synthesis of nitrogen-containing substances under conditions of microwave activation Lambert Academic Publishing, 2011. - 354 c.

12.   U.S. Pat. 4,539,338 USA. Phenol Formaldehyde Resoles For Making Phenolic

Foam/ E.W. Kifer, P.A.Trafford, V.J. Wojtyna; published 03/09/1985.

13. U.S. Pat. 31539463A USA. Foaming phenol-formaldehyde resins with fluorocarbons / Alessandro, William D. J.; published 06/18/1968.

14. U.S. Pat. 35773073A USA. PROCESS FOR CONTROLLED CURING OF FOAMS/ BUNCLARK E; published 05/07/1973.

15. U.S. Pat. 4033910A USA. Methyl formate as an adjuvant in phenolic foam formation / Papa A.J.; publ. 07/05/1977.

16. U.S. Pat. 4303758A USA. Method of preparing closed cell phenol-aldehyde foam and the closed cell foam thus prepared/ Gusmer, Frederick E.; publ. 12/01/1981.

17. U.S. Pat. 20080280787 A1 USA. Phenol-formaldehyde novolac resin having low concentration of free phenol/ Rediger, Richard A. Lucas, Edward; publ. 11/13/2008.

18. U.S. Pat. 20140287238 A1 USA. Phenol-formaldehyde novolac resin having low concentration of free phenol/ Rediger, Richard A. Lucas, Edward; published 09/25/2014.

19. U.S. Pat. 20120295114 A1 USA. Phenol-Formaldehyde Novolac Resin Having Low Concentration of Free Phenol / Rediger, Richard A. Lucas; published 11/22/2012.

yes

I want morebooks!

Buy your books fast and straightforward online - at one of world's fastest growing online book stores! Environmentally sound due to Print-on-Demand technologies.

Buy your books online at
www.morebooks.shop

Kaufen Sie Ihre Bücher schnell und unkompliziert online – auf einer der am schnellsten wachsenden Buchhandelsplattformen weltweit! Dank Print-On-Demand umwelt- und ressourcenschonend produzi ert.

Bücher schneller online kaufen
www.morebooks.shop

info@omniscriptum.com
www.omniscriptum.com

Printed by Books on Demand GmbH, Norderstedt / Germany